Diego Aparecido Carvalho Moreira
Diogo A.R. Gonçalves

Effect of nitrogen sources on bean plants using plant hormones

Diego Aparecido Carvalho Moreira
Diogo A.R. Gonçalves

Effect of nitrogen sources on bean plants using plant hormones

Field experiment

ScienciaScripts

Imprint

Any brand names and product names mentioned in this book are subject to trademark, brand or patent protection and are trademarks or registered trademarks of their respective holders. The use of brand names, product names, common names, trade names, product descriptions etc. even without a particular marking in this work is in no way to be construed to mean that such names may be regarded as unrestricted in respect of trademark and brand protection legislation and could thus be used by anyone.

Cover image: www.ingimage.com

This book is a translation from the original published under ISBN 978-613-9-65659-2.

Publisher:
Sciencia Scripts
is a trademark of
Dodo Books Indian Ocean Ltd. and OmniScriptum S.R.L publishing group

120 High Road, East Finchley, London, N2 9ED, United Kingdom
Str. Armeneasca 28/1, office 1, Chisinau MD-2012, Republic of Moldova, Europe
Printed at: see last page
ISBN: 978-620-7-79414-0

SUMMARY

SUMMARY

MOREIRA, D. A. C. **Effects of *Rhizobium tropici* bacteria and plant growth regulator on bean plant (*Phaseolus vulgaris*).** Araxâ, UniAraxâ, 2017. 42f.

The bean (*Phaseolus vulgaris) is* a legume belonging to the fabaceae family, widespread throughout the country and an important source of protein in human food. Inoculation with rhizobia has been widely discussed, based on studies that indicate its feasibility for use in agriculture, and various innovative research projects have been carried out, such as the use of plant growth regulators. The aim of this study was to elucidate issues relating to the use of growth regulators in bean cultivation, as well as the beneficial effects or otherwise of inoculation with *Rhizobium* bacteria. The experiment was carried out in the Uniaraxà Experimental Field, located in Araxà - MG. The experimental design used was randomized blocks in a 2x2 factorial scheme, with factor 1 being the source of nitrogen and factor 2 being the application of a growth regulator, in four blocks. The nitrogen sources in factor 1 were mineral fertilization using urea and inoculation using *Rhizobium tropici,* while factor 2 was whether or not the growth regulator was used. The cultivar used in the experiment was BRS Estilo, with a population of 240,000 plants ha^{-1}. The treatments were: T1 (50 kg ha^{-1} of mineral nitrogen), T2 (50 kg ha^{-1} of mineral nitrogen + plant growth regulator), T3 (inoculation with *Rhizobium tropici*) and T4 (inoculation with *Rhizobium tropici* + plant growth regulator). The following items were assessed: number of nodules per plant (no.); dry mass of the aerial part (g pl^{-1}); dry mass of the root (g pl^{-1}); number of pods (no.); weight of 100 grains (g) and yield (Kg ha^{-1}). Under the conditions of this study, the nitrogen sources provided similar results between the treatments for almost all the variables analyzed, except for: number of nodules and dry mass of the aerial part. In relation to the plant growth regulator, there were significant differences between treatments T1 and T2 for the following variables: number of pods, weight of 100 grains and productivity Kg ha^{-1}.

KEYWORDS: BNF, nitrogen source, symbiosis, stimulate.

1. INTRODUCTION

The common bean is classified as belonging to the order Rosales, family Fabaceae, subfamily Faboideae, tribe Phaseoleae, genus Phaseolus and species Phaseolus *vulgaris* (VILHORDO, 1996).

There are several hypotheses to explain the origin of cultivated and domesticated forms of the bean plant. Wild types, similar to sympatric creole varieties, found in Mexico and domesticated types dating back to around 7,000 BC in Mesoamerica, support the hypothesis that the bean plant was domesticated and transported to South America. Another hypothesis is that the bean plant was domesticated in South America and transported to North America, with older archaeological finds (10,000 BC) of domesticated beans in South America (Peru) being found than those in Mesoamerica. More current data based on electrophoretic patterns of phaseolin (the main reserve protein of the bean) indicate the existence of two primary centers of domestication, Mesoamerica and the southern Andes (ARAUJO et al., 1996).

The application of mineral nitrogen fertilizers to tropical soils has often resulted in a low frequency of response by the bean crop (VIEIRA, 1998). The low efficiency of chemical nitrogen fertilizers (around 50%) by plants is due to losses through leaching, volatilization and denitrification.

Biological nitrogen fixation (BNF) is of great importance in agriculture, partly reducing the environmental liabilities generated by the high consumption of nitrogen fertilizers, which not only consume a lot of fuel for their industrial production, there are other problems such as the contamination of water and food by NO_3^- and NO_2^- , plant toxicity due to the presence of high levels of NO_2^- in the soil, changes in plant development due to excess N and CO_2 emissions contributing to global warming (MARIN et al., 1999).

Beans develop a symbiotic association in their roots with the bacterium *Rhizobium tropici*. When this bacterium is present in the soil, naturally or via inoculation, it recognizes and infects the roots of the host plant, causing the formation of a nodule in the root system where it fixes atmospheric nitrogen (N_2) and transforms it into NH_3 and NH_4, partially or totally supplying the nitrogen required by the plant (HUNGRIA et al., 1997).

Rhizobium tropici stands out among the most efficient bacterial species for BNF in bean because it has greater genetic stability and is more tolerant of higher temperatures and soil acidity (MARTINEZ-ROMERO et al., 1991).

Regulators are a mixture of plant regulators (synthetic hormones) and nutritional substances, which in appropriate quantities are intended to modify and positively influence both the morphology and physiology of plants CASTRO & VIEIRA (2001).

Considering the importance of the bean crop in the daily diet of Brazilians, studies that can improve the agronomic techniques of this crop are of the utmost importance. This work seeks to elucidate issues related to the use of growth regulators in the bean crop as well as the beneficial effects or not of inoculation with bacteria of the genus *Rhizobium*.

2. OBJECTIVES

2.1 GENERAL

- To evaluate the effects of the *Rhizobium tropici bacterium* and the use of plant growth regulators on the bean plant (*Phaseolus vulgaris*).

2.2 SPECIFIC

- Count the number of nodules per plant (n°);

- Calculate the dry mass of the aerial part (g pl)$^{-1}$;

- Measure root dry mass (g pl)$^{-1}$;

- Evaluate the number of pods (n°);

- Measure the weight of 100 grains (g);

- Estimate productivity (Kg ha)$^{-1}$.

3. BACKGROUND

Nitrogen is the nutrient that the bean plant demands in the greatest quantities (FANCELI and DOURADO NETO, 2007), and in addition to the high economic cost, the use of nitrogen fertilizers in tropical soils also has an additional ecological cost. The losses of nitrogen fertilizers applied are considered to be around 50%, and are mainly caused by leaching, in the form of nitrate, and surface run-off, caused by rainwater and/or irrigation (STRALIOTTO et al., 2002). The N lost in this process is highly polluting and, once carried into the groundwater, causes contamination of underground aquifers, rivers and lakes. Other losses of applied N occur in gaseous forms, which return to the atmosphere, mainly through the processes of denitrification and volatilization (SIQUEIRA et al., 1994; STRALIOTTO et al., 2002). In this context, the proper management of nitrogen fertilization is one of the main difficulties facing the bean crop, since the application of excessive doses of N, in addition to increasing the economic cost, can pose serious risks to the environment, and its use in insufficient quantities can limit its productive potential, even if other production factors are optimized (SANTOS et al., 2003).

The high costs of nitrogen fertilizers have made it increasingly interesting to evaluate the biological nitrogen fixation (BNF) capacity of commercial bean cultivars, as BNF is a low-cost technology that can contribute significantly to reducing the production costs of this legume.

Given the current situation in Brazilian agriculture, there is a growing search for technologies that increase productivity with the least possible impact on the environment. With this in mind, inoculation with *Rhizobium tropici* bacteria and the use of plant growth regulators in the bean plant (*Phaseolus vulgaris*) have become alternative focuses for studies with crops of general agronomic interest.

4. THEORETICAL FRAMEWORK

4.1 Economic and regional aspects of bean cultivation

The common bean (*Phaseolus vulgaris*) is one of the main crops produced in Brazil and worldwide. Its importance goes beyond the economic aspect, due to its relevance as a factor in food and nutritional security and its cultural importance in the cuisine of various countries and cultures, being an important source of protein in the human diet. Currently, per capita consumption has shown a slight increase, and in 2010 was around 17.06 kg/inhabitant/year (BARBOSA & GONZAGA, 2012). Its production, processing and marketing chains generate jobs and income, especially for the less privileged classes (FACHINI et al., 2006).

The great advantage of this legume is its wide adaptation to different climates, which allows it to be grown all year round, in almost all Brazilian states, during different seasons and harvests (SALVADOR, 2011).

The main bean producing countries are Brazil, India, China, Myanmar and Mexico, which account for over 65% of world production. The main exporting countries are China, the USA, Myanmar, Canada and Argentina, which together account for 73.5% of total exports. The main bean importing countries are India, the USA, Cuba, Japan and the United Kingdom (WANDER et al., 2007).

[a]There are three sowing seasons for the bean crop: the 1st, or water bean season, with sowing between August and November and climatic conditions that are normally satisfactory for the crop, with regular rainfall and no need for irrigation; in the 2nd season or dry beans, sowing takes place from February to March with the advantage of harvesting during the dry season, but the lack of rainfall during the cycle results in lower yields; and finally, the 3rd season or winter beans, sowing in April and May, also with the advantage of harvesting during the dry season, but with the need for an irrigation system (CHAIM, 2011).

As a typical product of the Brazilian diet, beans are grown by small and large producers in all regions. The states with the highest production are Paraná, which harvested 298,000 tons in the 2009/2010 crop, and Minas Gerais, which produced 214,000 tons in the same period (MAPA, 2012). The crop has a projected annual increase rate of 1.77%, according to a study by the Ministry of Agriculture's Strategic Management Office. The data also shows a growth in consumption of around 1.22% per year, from 2009/2010 to 2019/2020, rising from 3.7 million tons to 4.31 million tons (MAPA, 2012).

The carioca type of bean is the most widely grown in Minas Gerais and Brazil. They are cream-colored with brown streaks. In recent years, breeding programs have obtained numerous cultivars with this type of grain. Most of them have advantages in terms of productivity and disease resistance over the original carioca (RAMALHO; ABREU, 1998).

Despite the country's low productivity of around 850 kg ha^{-1} , the bean crop has been exploited in various production systems, achieving yields of over 3,000 kg ha^{-1} (YOKOYAMA, 2002).

According to the results obtained by SANTOS (2012), the productivity achieved during the winter period in Uberlândia - MG, ranged from 2713.24 kg ha^{-1} to 3844.57 kg ha^{-1} where 32 genotypes of the Carioca group were evaluated, in which the Pérola cultivar was statistically among the lowest with a productivity of 3125.92 kg ha^{-1} .

In the dry season in Uberlândia - MG, according to SANTOS (2010), productivity varied from 335.43 kg ha^{-1} to 2073.58 kg ha^{-1} , where 26 genotypes of Carioca bean were evaluated, where the cultivar was statistically among the superior cultivars, with 1737.95 kg ha^{-1} .

COBUCCI et al. (2005), in three experiments carried out in Unai - MG, regardless of whether conventional or no-till cultivation was used, average bean yields ranged from 2148 to 3816 Kg ha^{-1} .

4.2 Bean Morphophysiology

The common bean (*Phaseolus vulgaris*) is an annual, diploid (2n=2x=22) and autogamous species, where self-pollination predominates (95%). The common bean has a short life cycle, lasting an average of 90 days depending on the cultivar and environmental conditions (ALMEIDA, LEITAO FILHO and MIYASAKA, 1971).

It is a herbaceous plant whose pivoting root system reaches approximately 20 cm in length, with most of it located in the top 10 cm of the soil. The stem is herbaceous. The leaves are petiolate, trifoliate and alternate, except for the first ones, which are simple and opposite. In trifoliate leaves, the central or terminal leaflet is symmetrical and acuminate and the lateral leaflets are asymmetrical and acuminate (LEON, 1968; VIEIRA, 1967).

There are two growth habits: determinate, inflorescence at the apex of the main and lateral stems, and indeterminate, where flowering occurs from the base to the apex (ALMEIDA, LEITAO FILHO and MIYASAKA, 1971).

The development of the bean plant basically comprises two distinct and successive phases, called the vegetative and reproductive phases, differentiated from each other by the manifestation of different biochemical, morphological and physiological events. The vegetative phase begins with the complete unfolding of the primary leaves and continues until the first flower buds appear (DOURADO NETO & FANCELLI, 2000a and FANCELLI, 1990a, b, 1992, 1994). The vegetative period is favored by moderately high temperatures (above 21.0°C and below 29.5°C), adequate water availability and abundant light. The reproductive phase lasts from the first flower buds to the point of physiological maturity.

The vegetative phase consists of stages V0, V1, V2, V3 and V4 and the reproductive phase of stages R5, R6, R7, R8 and R9 (LAING et al., 1984).

- VO - Germination (seed germination has started);

- V1 - Emergence (50% of the cotyledons out of the ground);

- V2 - Primary leaves (pair of expanded primary leaves);

- V3 - First trifoliolate leaf (with expanded leaflets);

- V4 - Third trifoliate leaf (with expanded leaflets);

- R5 - Pre-flowering (after the first bud or flower raceme has appeared);

- R6 - Flowering (first flower opened);

- R7 - Vegetable formation (first pod with corolla detached);

- R8 - Vegetable filling (pods begin to swell);

- R9 - Ripening (when the first pod begins to discolor or dry out).

Beans are classified into four main types according to their growth habit: Type I, comprising plants with determinate growth; Type II, characterized by indeterminate and bushy growth; Type III, indeterminate and prostrate growth and; Type IV, indeterminate and climbing growth (CIAT, 1978).

A bean plant is made up of apparently distinct parts. In the soil, there is a root system and above that, a stem that bears the leaves and branches. In older plants, you can get a detailed view of its parts: root, stem, leaves and axillary stems, inflorescences, fruit and seed (GAVILANNES, 1995).

The factors that affect bean productivity are the number of pods per plant, the number of kernels per pod and the mass of the kernels. These variables are important for selecting more productive genotypes (COSTA et al., 1983; COIMBRA et al., 1999).

4.3 Nitrogen sources

Nitrogen (N) is one of the mineral elements most required by plants and the one that most limits growth, but when properly fertilized, N is the most responsive nutrient. Its high response to nitrogen fertilization is due to its active participation in plant metabolism, as a constituent of proteins, nucleic acids and many other cellular constituents, membranes and various phytohormones (SOUZA; FERNANDES, 2006).

N is the most abundant element in the atmosphere, making up around 78% of atmospheric air, which represents its large natural reservoir. It is also present in soil, water and living beings. Even with this abundance in the environment, N is generally a limiting factor for plants, as only around 0.0007% of it is easily available to plants (HAVLIN, 2005).

The main sources of N supply to the crop are the soil, through the decomposition of organic matter, nitrogen fertilizers and the process of biological fixation of atmospheric nitrogen N_2, through the symbiosis of the bean plant with bacteria from the rhizobia group (HUNGRIA et al., 1997; MERCANTE et al., 1999).

The efficiency of nitrogen fertilization depends not only on doses, but also on the nitrogen sources used, and it is essential to combine N sources and doses with soil and crop management practices. After application, nitrogen fertilizers suffer various influences in the environment, which can lead to losses through leaching, volatilization and denitrification. When leaching occurs, nitrate is carried deep into the soil and can reach the water table, contaminating water sources (VILLAS BOAS et al., 1999).

Another significant loss of N is due to NH_3 volatilization, when urea is applied to the soil surface in unsuitable conditions, which can cause losses of 40 to 78% of the N applied to the soil surface when poorly managed (LARA CABEZAS et al., 2000).

Among N sources, the main fertilizers produced are synthesized from atmospheric N_2 and H from fossil fuels (due to their economic competitiveness when compared to other energy sources). This combination of N_2 + H results in a molecule of NH_3 (CANTARELLA, 2007), which is the basis for the production of other nitrogen fertilizers. It is estimated that between 1.2 and 1.8% of the world's fossil energy consumption is used to produce nitrogen fertilizers (LANGREID et al., 1999 cited by CANTARELLA, 2007).

Among the nitrogen fertilizers used in agriculture, urea stands out. It comes from the reaction of liquid NH_3 obtained synthetically with CO_2 in closed chambers, generating urea, ammonium carbamate and ammonia. This mixture is separated with the aid of a rotaevaporator, obtaining crystalline urea, $CO(NH_2)_2$ (MALAVOLTA, 1967). This simple mineral fertilizer has a solid physical characteristic (ALCARDE et al., 1998), with an N content ranging from 44 to 46%, which can be lower if some product is added, such as coatings or inhibitory substances. Another relevant characteristic for its use is the cost per kg of N. Among solid nitrogen fertilizers, urea has the lowest cost per kg of N. However, when poorly managed, it can cause major losses due to volatilization.

Ammonium nitrate and ammonium sulphate are two other important nitrogen fertilizers used in Brazilian agriculture, with N contents of 33% and 21%, respectively. The advantage of using ammonium nitrate is that it is not susceptible to loss by volatilization due to the action of urease, but there are strict controls on its trade, especially outside Brazil, due to its use in the manufacture of explosives.

Ammonium sulphate not only doesn't cause losses due to ammonia volatilization, but is also made up of parts of the sulphur molecule (22 to 24% S), but results in a lower concentration of N in its composition compared to urea and ammonium nitrate. Other fertilizers used are ammonium phosphates (DAP and MAP), but they are mostly used for phosphate application (CANTARELLA, 2007).

Beans develop a symbiotic association in their roots with the bacterium *Rhizobium tropici*. When this bacterium is present in the soil, naturally or via inoculation, it recognizes and infects the roots of the host plant, causing the formation of a nodule where N2 fixation occurs (HUNGRIA et al., 1997). Once the nodule has started, the Rhizobium transforms into a bacteriophage that multiplies and starts synthesizing nitrogenase, the enzyme responsible for reducing N2 and initiating N2 fixation (HUNGRIA et al., 1997).

Biological fixation is the most important way of fixing N2 into ammonia (NH_3) and is of unique importance in the biogeochemical cycle of this element. It is also an alternative for sustainable soil management (HUNGRIA; VARGAS, 2000). According to CARVALHO (2002), BNF has numerous advantages, including its low cost and, consequently, its accessibility to producers, the absence of environmental problems due to its lack of impact and the abundance of N in the atmosphere, making it an inexhaustible source. In this context, BARBOSA et al. (2010) mention that the soil is an easily exhaustible source of N due to successive crops.

4.4 Chemical nitrogen fertilization for bean plants

In Minas Gerais, official recommendations for the fertilization of common beans vary from 20 to 40 kg ha^{-1} of nitrogen at planting and from 20 to 60 kg ha^{-1} of nitrogen in top dressing, depending on the technological level of the producer and the expected productivity (CHAGAS et al., 1999).

For the bean crop, most nitrogen recommendations are aimed at the conventional tillage system. CARVALHO et al. (1992) recommend 90 kg ha^{-1} of N to achieve maximum productivity. SILVEIRA and

DAMASCENO (1993) recommended only 72 kg ha^{-1} of N to maximize bean productivity. SILVA et al. (2000), in a conventional tillage system, obtained a quadratic response of the bean plant to N and maximum productivity was achieved with 74 kg ha^{-1} of nitrogen.

4.5 Bean inoculation

With the discovery of the *Rhizobium tropici* species associated with the bean plant and the selection of the most efficient strains currently recommended, positive experimental results have been accumulating with inoculation.

The procedure for inoculating seeds with rhizobium is simple: just mix the seeds with rhizobium inoculant for beans. This inoculant is generally sold in packages containing the bacteria in a peat vehicle, which is currently the most recommended by research.

Studies have shown that it is possible for this crop to benefit, under field conditions, from the process of biological nitrogen fixation, and can achieve yields of over 2,500 kg ha^{-1} (HUNGRIA et al., 2000). However, it should be borne in mind that the success of bean inoculation with highly efficient rhizobium strains is associated with the competitive ability of these strains and their adaptation to environmental conditions (MERCANTE et al., 1999; STRALIOTTO et al., 2002). Under suitable environmental conditions, the atmospheric nitrogen fixed by the symbiosis can meet most of the bean plant's needs (HUNGRIA et al., 1985). However, the factors of soil acidity, low pH and high concentrations of toxic Al often limit all stages of the process of root infection, nodule formation and N assimilation by the plant (DERNADIN, 1991; MARTINEZ-ROMERO et al., 1991; MERCANTE, 1993).

When introducing selected rhizobia via inoculants, it is important to consider the possibility of naturalized populations present in the soil promoting nodulation, except when the inoculation takes place in sterilized soil conditions. Therefore, the rhizobium strains in the inoculant must not only be efficient in BNF, but also sufficiently competitive to be able to outperform the native ones, ensuring greater nodulation (MATOSO; KUSDRA, 2014).

4.6 Growth regulators

Growth regulators are synthetic chemical substances that have an effect on plant metabolism (LAMAS, 2001).

The use of plant regulators can increase plant growth and development by stimulating cell division, differentiation and cell elongation, and can also increase the absorption and use of water and nutrients by plants (VIEIRA & CASTRO 2004). According to HINOJOSA (2005), there are five types of hormones considered classic hormones: auxins, gibberellins, cytokinins, ethylene and inhibitors, which are synthesized in different places in the plant.

Auxin plays a fundamental role in a wide range of growth and development processes in plants. Taking the plant as a whole, auxin plays an important role in root formation, apical dominance, tropism, senescence and other processes (MERCIER, 2004). However, according to HINOJOSA (2005), the main characteristic of auxins is their ability to induce cell elongation.

Cytokinins derive their name from their ability to promote cell division or cytokinesis. They promote the growth of lateral buds and are also important for the maintenance and growth of the apical bud of plants (CID, 2005). Recent experimental work has shown that cytokinins protect cell membranes from peroxidation, which has been interpreted as a way of inhibiting leaf senescence (CID, 2005). They have a direct effect on plant growth and development. They work together with gibberellins to induce germination and enzymatic processes (RAVEN et al., 2001).

Gibberellins are plant hormones whose main activities are related to stem growth, the breaking of dormancy, the induction of parthenocarpy and other physiological processes. Gibberellic acid (GA_3), which is one of the most active gibberellins, has been commercially available for many years (MATSUMOTO, 2005). They control characteristics such as plant height, flowering time, fruit size, etc.

CASTRO et al. (1985) reported that immersing bean seeds in solutions containing plant regulators enables dormancy to be broken, uniform emergence and morphological and physiological changes in the seedlings, as well as avoiding phytotoxicity from these products when applied to the plant part, by using them pre-emergently. CASTRO et al. (1990) obtained greater growth and number of leaves in bean plants with gibberellin and reduced productivity with naphthalene acetic acid (ANA). HARB (1992) obtained increases in the number of leaves, fresh mass of roots, dry mass of seedlings, as well as the accumulation of nutrients, sugar and productivity of beans and cotton, by applying gibberellic acid to their seeds, with the greatest increases in the seed production of their crops being obtained with indole acetic acid (IAA).

4.7 Stimulate®

The plant growth regulator, commercial product Stimulate® from Stoller Interprises Inc., contains plant regulators and traces of chelated mineral salts. Its constituent plant regulators are as follows: indolebutyric acid (auxin) 0.005%, kinetin (cytokinin) 0.009% and gibberellic acid (gibberellin) 0.005%, This chemical product increases plant growth and development by stimulating cell division, differentiation and cell elongation, It also increases the absorption and use of nutrients and is especially effective when applied in seed treatments, applications with foliar fertilizers, and is also compatible with pesticides (STOLLER DO BRASIL, 1998; CASTRO et al., 1998; VIEIRA & CASTRO, 2004).

CASTRO & VIEIRA (2003) found that applying Stimulate® 10 mL L^{-1} to corn seeds increased seed germination. This same concentration increased the number of normal seedlings and consequently reduced the percentage of abnormal seedlings. VIEIRA (2001) noticed an increase in the growth parameters of the rice root system with the application of **Stimulate®** 4 mL kg^{-1} of seed. Concentrations of up to 10 mL kg^{-1} of seed increased the dry mass of the roots and aerial part. **Stimulate®** at 4 mL kg^{-1} of seed increased the number of panicles and the dry mass of rice grains.

CATO et al. (2004) studied the synergism between the growth regulators constituents of **Stimulate®** by applying them alone, in pairs or all three together to tomato plants. A mixture of the three pure components of the biostimulant at the concentrations of the commercial product was used as a control. The results showed that only the mixture of the three components increased tomato fruit yield.

VELLINI & ROSOLEM (1997), evaluating the agronomic efficacy of Stimulate® on bean plants, concluded that this product has a positive effect on productivity when applied in conjunction with the mineral nutrients cobalt and molybdenum, and can also increase protein production. OLIVEIRA et al. (1997), when applying 250 mL ha^{-1} of Stimulate® to bean seeds, recorded higher yields of pods and kernels per plant.

VIEIRA & CASTRO (2003) found that applying Stimulate® to bean seeds resulted in an increase in germination up to a concentration of 8 mL kg^{-1} of seed and at a concentration of up to 10 mL kg^{-1} of seed increased the germination of normal seedlings and decreased the emergence of abnormal bean seedlings. CASTRO et al. (2005) observed that Stimulate® increased the dry mass of bean roots up to a concentration of 10 mL kg^{-1} of seed. The biostimulant also increased the number of pods per plant, the number of kernels per pod and the dry mass of kernels per plant at a concentration of 5 mL kg^{-1} of seed.

4.8 Interaction between nitrogen sources and plant growth regulators

The growing use of cultivars with high production potential has meant more frequent use of inputs, including nitrogen. However, the use of increasingly high doses of this nutrient in order to increase productivity leads to high vegetative growth, which causes plant lodging and negatively affects productivity and grain quality. The problem of lodging can be minimized with the use of resistant cultivars, which has been occurring, and with the use of growth regulators, which in addition to reducing the size of the plant provide better use of nutrients, due to the physiological changes they exert on the plant (BUZETTI et al., 2006).

When comparing the effect of nitrate and ammonium on the auxin content in the roots and the growth of wheat seedlings (*Triticum durum*) at different temperatures, it was found that the length of the roots, the aerial part/root ratio and the auxin content were higher in the plants that received nitrate. The higher rate of root growth in the presence of nitrate and at higher temperatures was associated with a temporary increase in auxin concentration (KUDOYAROVA et al., 1997).

SRIVASTAVA et al. (1994), evaluating the physiological responses of bean plants to nitrate, observed that cytokinin accelerates the absorption of this nutrient and also prevents damage that may occur due to excess, as well as accumulating N in the tissues. In this sense, some studies have indicated that the application of phytoregulators favors the absorption and assimilation of N by plants, leading to greater productivity and protein content (CATE; BRETELER, 1982; CHANDA et al., 1998; RUIZ et al, 2000).

BARBOSA et al. (2008), evaluating bean seed production as a function of the application of the biostimulant Stimulate® associated with different doses of urea, found significant results for the percentage of normal pods, height of insertion of the first pod, number of seeds per pod and mass of 100 grains, when they used the doses 0.0; 0.5; 1.0; 1.5 and 2.0 L ha^{-1} of Stimulate® .

5. MATERIAL AND METHODS

The experiment was conducted in the Experimental Field of the Centro Universitário do Planalto de Araxà - UNIARAXA, located in the municipality of Araxà MG, at 19°34'45.2"S and 46°57'15.3"W, at an altitude of 932 m, on a medium-textured, eutrophic red latosol. The region's climate, according to the Koppen classification, is cwa (humid temperate climate with dry winters and hot summers).

The experimental design used was randomized blocks in a 2x2 factorial scheme, with factor 1 being the source of nitrogen and factor 2 being the application of a growth regulator, with four blocks. The nitrogen sources were mineral fertilization using urea and inoculation using *Rhizobium tropici,* while factor 2 was whether or not the growth regulator was used. The experimental plots were sized at 3.0m x 3.0m. Totaling 9m^2 per plot. Each plot was made up of six 3.0 m long lines, spaced 0.50 m apart and 12 plants per linear meter, giving a total stand of 240,000 plants per hectare.

Six simple soil samples were taken at a depth of 020 cm to form a composite sample (a mixture of all the other 6 simple samples) from the experimental area and determine the chemical analysis, the values of which are described in figure 1 below.

FIGURE 1. Chemical analysis of the soil

pH	mg/dm^3		Cmolc/dm3							%	
H2O	P	K	Al	Ca	Mg	H+Al	SB	t	T	V	M
6,26	21,97	209,0	0,02	1,92	0,66	2,15	3,11	3,13	5,26	59,21	0,64

Source: Profert-MG Laboratory

The planting fertilizer was determined on the basis of the crop's needs, complementing the nutrient levels already present in the soil. Therefore, 70 kg ha^{-1} of P2O5 and 20 kg ha^{-1} of K2O were applied to all treatments, both treatments received nitrogen fertilizer at planting at a dose of 20 kg ha^{-1} of N, while only treatments 1 and 2 received chemical nitrogen at a dose of 50 kg of N ha^{-1} , divided between 20 and 30 days after emergence (DAE), in two top dressings, 56.81 kg ha^{-1} of urea, per top dressing.

The inoculant used for the work was a peat-based inoculant produced by NITRO1000 Ind. e Com. de Prod. Agropecuârios e Tex. Ltda., recommended for growing beans, containing *Rhizobium tropici* strains (Semia 4077/Semia 4088), with a bacterial concentration of 3.0x109 CFU/g, using the dose recommended by the manufacturer, which is 2 g kg^{-1} of seed.

The plant growth regulator used was the commercial product Stimulate® from Stoller Interprises Inc., using the dose recommended by the manufacturer for seed treatment, which is 7.5 ml kg^{-1} of seed.

This resulted in the following treatments, as shown in Table 1 below:

Table 1. Description of treatments

TREATMENT	DESCRIPTION
T1	50 Kg ha^{-1} of mineral nitrogen
T2	50 kg ha^{-1} mineral nitrogen + RCV*
T3	Inoculation with *Rhizobium tropici*
T4	Inoculation with *Rhizobium tropici* + RCV

*RCV: Plant growth regulator

Sowing took place on October 29, 2016. The cultivar used in the experiment was BRS Estilo, a common bean cultivar of the carioca commercial group with an upright plant architecture, adapted to direct mechanical harvesting. It has high production potential and production stability. In terms of diseases, BRS Estilo is moderately resistant to anthracnose and rust, has intermediate resistance to common bacterial blight and rust and is susceptible to angular spot, golden mosaic and fusarium wilt.

At stage R6 of the crop (flowering), the following variables were evaluated: a) number of nodules per plant (no.), b) dry mass of the aerial part (g pl^{-1}); c) dry mass of the root (g pl^{-1}). At harvest stage R9 (when the first pod begins to discolor or dry out), the following variables were evaluated: d) number of pods (no.); e) weight of 100 grains (g); f) grain yield kg ha .$^{-1}$

To count the number of nodules, 5 plants were collected in a block with the help of a shovel per treatment, and a block of soil was removed along with the roots. After counting, the process of obtaining the values for the dry mass of the aerial part and the dry mass of the root began, where they were packed in duly identified paper bags and placed in an oven for 24 hours at a temperature of 100° Celsius. Afterwards, the dry mass of the aerial part and the dry mass of the root were weighed individually on a precision scale.

To determine the number of pods (n°), 10 plants were collected from the useful area of each plot. After counting, the pods were threshed to determine the weight of a hundred grains, using the average of three samples of 100 grains per experimental plot. To calculate grain yield, the entire useful area of the experimental plots was harvested and the grain weight obtained was extrapolated to kg ha^{-1} (13% moisture).

Weed control was carried out in two applications, one at stage v3 of the bean plant and the second at stage v4, using mixtures of selective non-systemic herbicides from the chemical group benzothiadiazinone and diphenyl ether, following the dosage recommended on the product's leaflet. A backpack pump was used to apply the herbicide.

In order to control insects, contact and ingestion insecticides from the pyrazole analog chemical group were applied. Two applications were made, one at stage v3 of the bean plant and the second 15 days after the first application. Using the dosage recommended by the manufacturer, a backpack pump was used for the application.

In order to control fungal diseases, contact fungicides from the alkylenebis chemical group were applied. Two applications were made, one at stage V3 of the bean plant and the second at stage R6. Using the dosage recommended by the manufacturer, a backpack pump was used for the application.

6. RESULTS AND DISCUSSION

For the Shapiro-Wilk normality test, we can see in figure 2 that the data presented for root dry mass did not meet the assumption of normality, with a distribution of data that was not reliable enough to carry out the mean comparison tests, This can be explained by the evaluation methodology used, since the roots were collected from the plants in the field, where despite the evaluation error being standardized for the whole, the collection was carried out by more than one person, which disharmonized the error, making it impossible to carry out the analysis of variance.

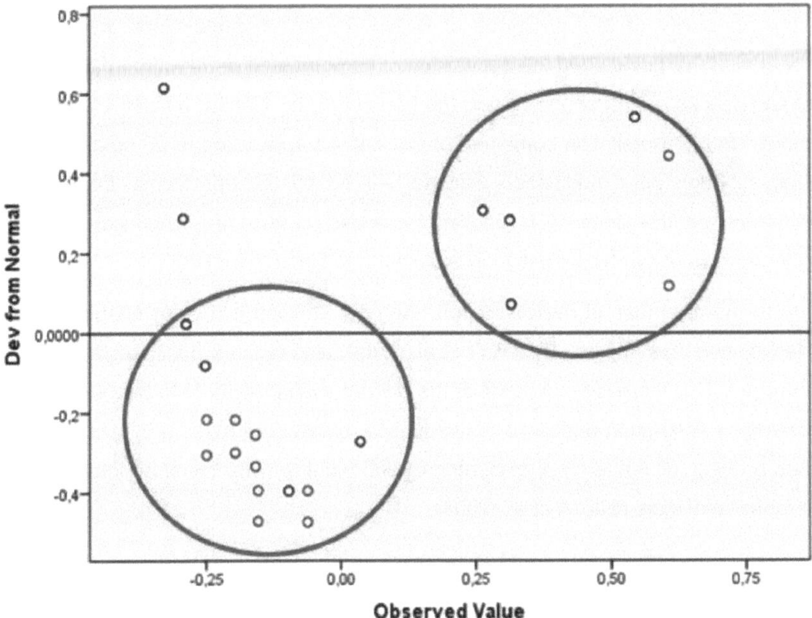

FIGURE 2 Scatterplots of normality of root dry mass by the Shapiro-Wilk test at 0.05 probability.

For the dry mass of the aerial part, the homogeneity of variances test accepted HO at a significance level of 0.05, i.e. it can be considered that the variances of the groups are different (^{1} F (p) = 5.987 (0.003)). The assumption of homogeneity of variances is thus rejected, so the Snedecor *F-test* cannot be used to assess the difference between mean values. According to ALMEIDA et al. (2008), Levene's test is used to compare the variances of groups of observations from continuous and not necessarily normal distributions. Levene's test is robust to non-normality, although some shortcomings have been highlighted by some authors, who have also presented more efficient alternatives.

The F-test obtained through one-factor analysis of variance to compare the means of independent normal populations deviates in terms of the size of the test when the groups have different population variances.

From the data presented in Table 2, as expected, the number of nodules was higher with inoculation, as opposed to chemical fertilization without inoculant, the bean cultivar used has already been highlighted in other works as having a good response to nodulation and N fixation (DIAS, 2017). The presence of nodules was also found in the chemical treatment without inoculant, indicating the occurrence of native strains in the soil, since nodulation was low, FULLIN et al. (1999) and VIEIRA et al. (2005) demonstrated in their research the possibility of this presence of native *Rhizobium* populations in the soil. PELEGRIN et al. (2009) evaluated the application of nitrogen fertilizer in different doses (0, 20, 40, 80 and 160 kg ha^{-1} of N as urea) in the bean crop, *cv.* Pérola, as well as controls with *Rhizobium tropici* inoculation combined or not with an application of 20 kg ha^{-1} of N, showed that although plant nodulation was not altered by the treatments, there were downward trends as the dose of N applied increased, due to the chemical fertilizer inhibiting the process of symbiosis between the plant and bacteria for BNF.

With regard to the dry mass of the aerial part, the data showed a greater quantity when chemical fertilization was applied. BINOTTI et al. (2009) also found a positive effect when applying nitrogen. This increase is a result of the greater availability of N to the bean plant, thus increasing its absorption and, as a consequence, greater dry mass production, since nitrogen has a direct influence on photosynthesis and plant growth, being an integral part of the chlorophyll molecule (SILVA et al., 2004).

Table 2. Influence of growth regulators in nitrogen nutrition, via chemical and biological means, on morphological and productive characteristics.

N source	Regulator	Number of nodules (n°)	Aerial part dry mass (g pl)$^{-1}$	Root dry mass (g pl)$^{-1}$	Number of pods (n°)	Weight of 100 grams (g)	Productivity (Kg ha)$^{-1}$
Chemistry	+	4.25 bA	6.09 aA	1.17 aA	14.25 aB	15.66 aB	2031.12 aB
Chemistry	−	9.25 bA	6.99 aA	1.23 aA	17.62 aA	16.28 aA	2844.31 aA
Inoculation	+	38.00 aA	4.00 bA	0.93 aA	15.62 aA	16.26 aA	2124.20 aA
Inoculation	−	41.25 aA	4.43 bA	1.14 aA	17.50 aA	16.38 aA	2742.60 aA
CV%		24,70 [1]	21,12 [2]	33,02	16,45	3,57	2.85 [3] 0.94 0.12* 0.77

W	0,91	0,90 (0,03)**	0,91 0,00^{ns}	0,93 0,18*	0,93 0,02**	(0.52)*
[1]F	(0,02)**	5,99 (0.00)^{ns}	1,17 (0.34)*	1,06 (0.38)*	2,96 (0.05)*	
	2,37					
	(0.09)*					

Lowercase letters express significance between N sources and uppercase letters represent significant variations between treatments with a specific regulator for each N source, using Tukey's test at 0.05 probability.

1 = Coefficient of variation for data transformed into ARCSEN √(X/100); 2 = Coefficient of variation for data transformed from X = 1/X; 3 = Coefficient of variation for data transformed from X=LOG(X)

However, the data on the dry mass of the aerial part was not homogeneous and did not meet this assumption, which makes it difficult to be sure of the reliability of the data, but even though it was not homogeneous, the data is still comparable with that of others, such as PELEGRIN et al. (2009), SILVA et al. (2004) and BINOTTI et al. (2009), who also confirm this greater development of the dry mass of the aerial part when chemical fertilization is carried out in view of biological nitrogen fixation.

With regard to root dry mass, it has already been explained that the data was not normal, due to the fact that the roots were collected in the field by different collectors, which may have interfered with standardizing the error, which would provide normal data. CASTRO et al. (2005) observed that Stimulate® increased the dry mass of bean roots, up to a concentration of 10 mL kg^{-1} of seed and VIEIRA (2001) testing on rice, obtained significant increases for this parameter. The higher the auxin concentration, the higher the growth rate of the root system according to KUDOYAROVA; FARKHUTDINOV; VESELOV (1997) and indolebutyric acid has the ability to promote the formation of root primordia, which is why it has been used to cause and accelerate the rooting of cuttings in the vegetative propagation of numerous plant species, i.e. this regulator is used to promote rhizogenesis (COLL et al., 2001).

With regard to the number of pods, inoculation and chemical fertilization did not show any differences, but when chemical fertilization was combined with the use of the plant growth regulator, there was a greater quantity per plant, and the application of the plant growth regulator was significant when combined with chemical fertilization. In the soybean crop, BERTOLIN et al. (2010) observed that Stimulate® provided an increase in the number of pods per plant and in soybean grain yield, both when applied via seeds and via foliar application and OLIVEIRA et al. (1998), evaluating the application of Stimulate® at different times, associated or not with the application of the product to bean seeds, found a significant effect on the number of pods per plant only when seed treatment was carried out, but when foliar application was carried out, they were unable to associate the effect of gibberellin or cytokinin with the production of bean pods.

In the variable, weight of 100 grains, which is also a productivity characteristic, the weight of the grains was higher when hormones were supplied, which helps the plant to develop better under conditions of chemical fertilization, which has a better acceleration of the plant's metabolism compared to inoculation. RODRIGO & FIOREZE (2015), analyzing the effect of regulators applied with chemical fertilization or fixing bacteria, report that the bioregulatory effect provided by symbiosis can replace those provided by the application of commercial stimulants or even act synergistically.

With regard to productivity in Kg ha^{-1} , this, in line with the other production data, also showed a difference only when chemical fertilization was used in conjunction with the plant growth regulator, while inoculation showed no difference when using the regulator or not, which is already a concept of the application of products such as growth regulators, as LEITE et al. (2003) mentions, when gibberellic acid is applied, it induces intense vegetative growth in various crops, which is greater than what is necessary for maximum productivity. As inoculation is a biological process aimed at development with greater health, it does not provide conditions for greater productive potential for the plant, while chemical fertilization provides an environment conducive to greater expression of its productive potential, and when combined with the plant growth regulator, this potential can be better expressed with chemical fertilization, thus showing that chemical fertilization with plant growth regulator produced 813.19 kg ha^{-1} more than when no regulator was applied, but chemical fertilization and inoculation in relation to productivity, there were no differences in the evaluation between nitrogen sources. These results are similar to those obtained by MORAES et al. (2010), and indicate the possibility of growing beans without the need to apply nitrogen fertilizers. By adopting this practice, it will be possible to reduce production costs and increase farmers' incomes, which is in line with PELIGRIN et al. (2009) who concluded that fertilizing with 20 kg ha^{-1} of N, plus inoculant with the *R. tropici CIAT 899* strain, made it possible to grow bean crops without the need for nitrogen fertilizers. *tropici strain* CIAT 899 made it possible to obtain grain yields in the bean crop equivalent to applying up to 160 kg ha^{-1} of N and that inoculation, plus fertilization with 20 kg ha^{-1} of N at planting, resulted in an increase in net revenue similar to the application of 160 kg ha^{-1} of N and higher than the treatment with the fertilization of 20 kg ha^{-1} of N, without the use of the inoculant, demonstrating its importance for obtaining greater profitability in bean cultivation.

7. FINAL CONSIDERATIONS

Under the conditions of this study, bean inoculation and mineral nitrogen fertilization gave similar results between the treatments for the variables analyzed, except for: number of nodules and dry mass of the aerial part.

In relation to the plant growth regulator, there were significant differences between treatments T1 and T2 for the following variables: number of pods, weight of 100 grains and productivity in Kg ha $.^{-1}$

8. REFERENCES

ALCARDE, J. C. GUIDOLIN, J. A. LOPES, A. S. **Os adubos e a eficiência das aduçôes**. 3. ed. Sâo Paulo: ANDA, 1998. 35 p. (Technical Bulletin, 3).

ALMEIDA, A. et al. **Modifications and alternatives to Levene's and Brown and Forsythe's tests for equality of variances and means.** *Rev.Colomb.Estad.* [online]. 2008, vol.31, n.2, pp.241-260. ISSN 0120-1751.

ALMEIDA, L. D.; LEITAO FILHO, H. F.; MIYASAKA, S. Caracteristica do feijâo carioca no cultivar: um novo cultivar. **Bragantia**, v. 30, p. 33-38, 1971.

ARAUJO, F.F; RAVA, C. A.; STONE, L. F.; ZIMMERMANN, M. J. Initiation of nodulation in seven cultivars of common bean inoculated with Rhizobium strains. **Pesq. Agropecu. Brasilia,** v. 31, n. 6, p. 435-443, 1996.

BARBOSA, F. R.; GONZAGA, A. C. O. **Technical information for the cultivation of common bean in the Central Brazilian Region: 2012-2014**. Embrapa Rice and Beans. Santo Antônio de Goiàs, GO, 2012.

BARBOSA, G. F.; ARF, O.; NASCIMENTO, M. S. do; BUZETTI, S. FREDDI, O. da S. Nitrogen in top dressing and foliar molybdenum in winter bean. **Acta Scientiarum Agronomy,** Maringà, v. 32, n. 1, p. 117-123, jan./mar. 2010.

BARBOSA, R. et al. Bean seed production as a function of biostimulant application and urea doses. In: CONGRESSO NACIONAL DE PESQUISA DE FEIJAO. CAMPINAS, 9., 2008, Campinas. **Proceedings... Campinas**: Agronomic Institute of Campinas, 2008. 1 CD-ROM.

BERTOLIN, D. C. et al. Increasing soybean productivity with the application of biostimulants. *Bragantia,* **Campinas**, v. 69, n. 2, p. 339-347, 2010.

BINOTTI, Flâvio Ferreira da Silva. **Nitrogen management in winter beans in succession to corn and Brachiaria in a no-till system.** 2009. 178f. Thesis (Doctorate) - Faculty of Engineering, Universidade Estadual Paulista-UNESP, Ilha Solteira, 2009.

BUZETTI, S.; BAZANINI, G.C.; FREITAS, J.G.; ANDREOTTI, M.; ARF, O.; SA, M.E.; MEIRA, F.A. Response of rice cultivars to doses of nitrogen and the growth regulator chlormequat chloride. **Pesquisa Agropecuâria Brasileira,** Brasilia, v.41, n.12, p.1731-

1737, 2006.

CANTARELLA, H. Nitrogen. In: NOVAIS, R. F.; ALVAREZ, V. V. H.; BARROS, N. F.; FONTES, R. L. F.; CANTARUTTI, R. B.; NEVES, J. C. L. (Ed). Soil Fertility. Viçosa, MG: **Sociedade Brasileira de Ciência do Solo**, 2007. p. 375-470.

CARVALHO, A. M.; SILVA, A. M.; COSTA, E. F.; COUTO, L. Influence of fertigation on grain yield and production components of common bean (*Phaseolus vulgaris* L.) cv. Carioca. **Ciência e Pràtica**, v. 16, n. 4, p. 503511, 1992.

CARVALHO, E. A. de. **Agronomic evaluation of nitrogen availability to the bean crop under a direct sowing system.** 2002. 63 f. Thesis (Doctorate in Agronomy) - Escola Superior de Agricultura "Luiz de Queiroz", Universidade de Sao Paulo, Piracicaba, 2002.

CASTRO, P. R. C.; APPEZZATTO, B.; LARA, C. W. A. R.; PELESSARI, A.;

PEREIRA, M.; MEDINA, M. J. A.; BOLONHESI, A. C.; SILVEIRA, J. A. G. Action of plant regulators on the development, nutritional and anatomical aspects and productivity of bean (*Phaseolus vulgaris*) cv. carioca. **Anais da ESALQ**, Piracicaba, v. 47, n. 1, p. 11-28, 1990.

CASTRO, P. R. C.; CATO, S. C.; VIEIRA, E. L. Bioregulators and biostimulants in bean plants. In: FANCELLI, A. L.; DOURADO-NETO, D. (Ed.). **Irrigated beans:** technology & production. Piracicaba: ESALQ, 2005. p. 54-62.

CASTRO, P. R. C.; GONÇALVES, M. B.; DEMÉTRIO, C. G. B. Effect of plant regulators on seed germination. **Anais da ESALQ**, Piracicaba, v. 2, p. 449-468, 1985.

CASTRO, P. R. C., PACHECO, A. C. E.; MEDINA, C. L. Effects of *Citrus* Micro Stimulate on the vegetative development and productivity of pear orange trees (*Citrus Sinensis* L. Osbeck). **Scientia Agricola**, Piracicaba, v. 55, n. 2, p. 338-341, 1998.

CASTRO, P.R.C.; VIEIRA, E.L. **Aplicaçao de Reguladores Vegetais na Agricultura Tropical.** 2001, 131p.

CASTRO, P. R. C.; VIEIRA, E. L. Bioregulators and biostimulants in corn cultivation. In: FANCELLI, A. L.; DOURADO-NETO, D. (Ed.). **Maize:** management strategies for high productivity. Piracicaba: ESALQ, 2003. p. 99-115.

CATE, C. H. H.; BRETELER, H. Effect of plant growth regulators on nitrate utilization by roots of nitrogen-depleted dwarf bean. **Journal Experimental Botany**, Oxford, v. 33, n. 1,

p. 37-46, 1982.

CATO, S. C.; CASTRO, P. R. C.; ONGARELLI, M. G.; CARVALHO, R. F.; PERES, L. E. P. Study of the synergism between auxins, gibberellins and cytokinins in the vegetative development and fruiting of *Lycopersicon esculentum* Mill. Cv. Micro-Tom. In: CONGRESSO BRASILEIRO DE FISIOLOGIA VEGETAL, 10., 2004, Recife. **Proceedings...** Recife, 2004.1 CD-ROM.

CHAGAS, J.M. et al. Beans. In: COMISSÂO DE FERTILIDADE DO SOLO DO ESTADO DE MINAS GERAIS. **Recommendations for the use of correctives and fertilizers in Minas Gerais:** 5ª approximation. Viçosa, MG. 1999. p. 306-307.

COBUCCI, T.; WRUCK, F. J.; SILVA , J. G. Response of bean plants (*Phaseolus vulgaris* L.) to applications of biostimulants and nutrient complexes. In: CONGRESSO NACIONAL DE PESQUISA DE FEIJÂO, 8., 2005, Goiânia.

Abstracts... Santo Antônio de Goiàs: Embrapa Arroz e Feijao, 2005. V. 2, p. 10781081.

COIMBRA, J.L.M. et al. Adaptability and stability of phenotypes in genotypes of colored bean (Phaseolus vulgaris.L.) in three different environments. **Ciência Rural**, Santa Maria, v.29, n.3, p.441-448, 1999

COLL, J. B.; RODRIGO, G. N.; GARCIA, B. S.; TAMES, R. S. **Fisiologia vegetal**. Madrid: Ediciones Piràmide, 2001. 566 p.

COSTA, J.G.C.; KOHASHI-SHIBATA, J.; COLIN, S.M. Plasticity in the common bean. **Pesquisa Agropecuària Brasileira**, Brasilia, DF., v.18, p.159-l67, 1983.

CHAIM, S.G. Genotypes of common bean, carioca group, in the water season, in Uberlândia-MG, 2011.25 f. **Monograph** (Graduation) - Institute of Agricultural Sciences, Federal University of Uberlândia, Uberlândia.

CHANDA, S. V. et al. Influence of plant growth regulators on some enzymes of nitrogen assimilation in mustard seedlings. **Journal of Plant Nutrition**, Philadelphia, v. 21, n. 8, p. 1765-1777, 1998.

CIAT. **Annual report 1977.** Cali, Colombia, 1978.

CID, L. P. B. Cytokinins in higher plants: synthesis and physiological properties. In: CID, L. P. B. Plant **hormones in higher plants**. Brasilia: Embrapa Recursos Genéticos e

Biotecnologia, 2005. p. 58-79

DERNADIN, N.D. **Selection of *Rhizobium leguminosarum* bv. *phaseoli* strains tolerant to acidity factors and resistant to antibiotics**. Piracicaba, Luiz de Queiroz College of Agriculture, 1991. 89p. (Master's thesis).

DIAS, P. A. S. **Genetic potential of elite lines of common bean for biological nitrogen fixation.** 2017. 111 f. Thesis (PhD in Genetics and Plant Breeding) - School of Agronomy, Federal University of Goiás, Goiânia, 2017.

DOURADO NETO, D.; FANCELLI, A. L. **Bean production**. Guaiba: Agropecuària, 2000a. ch.1, p.23-48: Ecophysiology and phenology.

FACHINI, C.; BARROS, V.L.N.P.; RAMOS JUNIOR, E.U.; ITO, M.A.; CASTRO, J.L. Importance of beans in Brazilian agribusiness. In: **Summaries of the 22nd Bean Field Day.** Capao Bonito, p.1-7, 2006.

FANCELLI, A. L. **The cultivation of irrigated beans**. Piracicaba: FEALQ; ESALQ, Department of Agriculture, 1990a. p.1-24: Basic aspects of bean physiology.

FANCELI, A. L.; DOURADO NETO, D. **Bean production**. Piracicaba: The authors. 2007. 386p.

FANCELLI, A. L. **Irrigated beans**. Piracicaba: FEALQ; ESALQ, Department of Agriculture, 1992. p.5-22: Phenology and climatic requirements of the bean plant.

FANCELLI, A. L. **Technology for bean production**. Piracicaba: SEBRAE, 1994. 154p.

FULLIN, E. A.; ZANGRANDE, M. B.; LANI, J. A.; MENDONÇA, L. F.; DESSAUNE FILHO, N. Nitrogen and molybdenum in the fertilization of irrigated bean crops. **Pesquisa Agropecuària Brasileira**, v. 34, n 7, p. 1145-1149, 1999.

GAVILANES, M. L. **Notas complementares ao texto da disciplina histologia e anatomia vegetal.** Lavras: UFLA, 1995. 51 p.

HARB, E. Z. Effect of soaking seeds in some growth regulators and micronutrients on growth, some chemical constituents and yield of faba beans and cotton plants, **Bulletin of Faculty of Agriculture-University of Cairo**, Giza, v. 43, n. 1, p. 429452, 1992. Supplement.

HAVLIN, J. L.; TISDALE, S. L.; BEATON, J. D.; NELSON, W. L. **Soil fertility and fertilizers:** an introduction to nutrient management. 7. ed. New Jersey: Pearson, 2005. 515 p.

HINOJOSA, G. F. Auxin in higher plants: synthesis and physiological properties. In: CID, L. P. B. **Vegetative hormones in higher plants**. Brasilia: Embrapa Recursos Genéticos e Biotecnologia, 2005. p. 15-57.

HUNGRIA, M.; ANDRADE, D.S.; CHUEIRE, L.M.O.; PROBANZA, A.; GUTIERREZ-MANERO, F.J. & MEGIAS, M. Isolation and characterization of new efficient and competitive bean (*Phaseolus vulgaris* L.) rhizobia from Brazil. **Soil Biol. Biochem.**, 32:1515-1528, 2000.

HUNGRIA, M.; NEVES, M.C.P. & VICTORIA, R.L. Nitrogen assimilation by bean plants; II. Absorption and translocation of mineral N and fixed N2. **R. Bras. Ci. Solo,** 9:202-209, 1985.

HUNGRIA, M.; VARGAS, M.A.T. & ARAUJO, R.S. Biological nitrogen fixation in bean plants. In: VARGAS, M.A.T. & HUNGRIA, M., eds. **Biology of cerrado soils**. Planaltina, Embrapa-CPAC, 1997. p.189-294.

HUNGRIA, M.; VARGAS, M. A. T. Environmental factors affecting N2 fixation in grain legumes in the tropics, with an emphasis on Brazil. **Field Crops Research,** Netherlands, v. 65, n. 2-3, p.151-164, mar. 2000.

KUDOYAROVA, G. R.; FARLHUTDINOV, R. G.; VESELOV, S. Y. U. Comparison of the effects of nitrate and ammonium forms of nitrogen on auxin content in roots and the growth of plants under different temperatures. **Plant Growth Regulation**, v. 23, n. 3, p. 207-208, 1997.

LAING, D. R.; JONES, P. G.; DAVIS, H. G. Common Bean (Phaseolus vulgaris L.). In: GOLDSWORTH, P. R.; FISHER, N. M. **The physiology of tropical field crops**. New York: John Willey, 1984. p.305-351.

LAMAS, F. M. Growth Regulators. In: Embrapa Agropecuària Oeste. **Cotton**: production technology. Embrapa Agropecuària Oeste; Embrapa Cotton, Dourados, 2001. 296p.

LARA CABEZAS, W. A. R.; TRIVELIN, P. C. O.; KORNDORFER, G. H.; PEREIRA, S. Balance of solid and fluid nitrogen fertilization of cover crops of maize in a no-till system in the triangulo mineiro. **Revista Brasileira de Ciência do Solo**, Viçosa, v. 24, n. 3, p. 363-376, 2000.

LEITE, V. M.; ROSOLEM, C. A.; RODRIGUES, J. D. Gibberellin and cytokinin effects on soybean growth. **Scientia Agricola,** Piracicaba, v. 60, n. 3, p. 537-541, 2003.

LEON, J. **Fundamentos botânicos de los cultivos tropicales**. San José: IICA, 487 p. 1968.

MALAVOLTA, E. Manual de quimica agricola: adubos E aduçâo. Sao Paulo: **Agronômica Ceres**, 1967. 606 p.

MARIN, V. A.; BALDANI, V. L. D.; TEIXEIRA, K. R. S.; BALDANI, J. I. **Biological nitrogen fixation**: nitrogen-fixing bacteria **of** importance for tropical agriculture. Seropédica: Embrapa-CNPAB, 1999. 24 p. (Embrapa-CNPAB. Documentos, 91).

MARTiNEZ-ROMERO, E.; SEGOVIA, E.; MERCANTE, F.M.; FRANCO, A.A.; GRAHAM, P. H. & PARDO, M.A. *Rhizobium tropici*, a novel species nodulating *Phaseolus vulgaris* L. beans and *Leucaena* sp. trees. Int. **J. Syst.**

Bacteriol., 41:417-426, 1991.

MATOSO, S. C. G.; KUSDRA, J. F. Bean nodulation and growth in response to the application of molybdenum and rhizobial inoculant. **Revista Brasileira Engenharia. Agricola Ambiental**, Campina Grande, v. 18, n. 6, p. 567-573. jun. 2014.

MATSUMOTO, K. Gibberellins in higher plants: synthesis and physiological properties. In: CID, L.P.B. **Vegetal hormones in higher plants**. Brasilia: Embrapa Recursos Genéticos e Biotecnologia, 2005. p. 80-101.

MERCANTE, F.M.; TEIXEIRA, M.G.; ABBOUD, A.C.S. & FRANCO, A.A. Biotechnological advances in bean cultivation under symbiotic conditions. **R. Univ. Rural**: Sér. Ciênc. Vida, 21:127-146, 1999.

MERCANTE, F.M. **Use of *Leucaena leucocephala* in obtaining**

of *Rhizobium* tolerant to high temperature for bean inoculation. Seropédica, Federal Rural University of Rio de Janeiro, 1993. 149p. (Master's thesis).

MERCIER, H. Auxins. In: KERBAUY, G. B. **Fisologia Vegetal**. Sao Paulo: Editora, 2004.

p. 217-249.

MINISTRY OF AGRICULTURE, LIVESTOCK AND SUPPLY. **Beans**.

Available at: <http://www.agricultura.gov.br/vegetal/culturas/feijao>. Accessed on: 22 Nov. 2012.

MORAES, W. B.; MARTINS FILHO, S.; GARCIA, G. de O.; CAETANO, S. de P.;

MORAES, W. B.; COSMI, F. C. Evaluation of biological nitrogen fixation in drought-tolerant bean genotypes. **IDESIA**, Arica, v. 28, n. 1, p. 61-68, 2010.

OLIVEIRA, R. F. de; PACE, L.; ROSOLEM, C. A. Bean production and nutritional status as a function of the application of a growth promoter. **Cientifica**, Sao Paulo, v. 26, n. 1/2, p. 203-212, 1998.

OLIVEIRA, R. F.; PACE, L.; ROSOLEM, C. A. **Bean production and nutrition as a function of the application of a growth promoter.** Botucatu: UNESP, Dept. Agricultura e Melhoramento Vegetal, 1997. 10 p. (Technical Report).

PELEGRIN, R. et al. Bean crop response to nitrogen fertilization and rhizobium inoculation. **Revista Brasileira de Ciência do Solo**, Viçosa, MG, v. 33, n. 1, p. 219-226, feb. 2009.

RAMALHO, M. A. P.; ABREU, A. de F. B. Cultivars. In: VIEIRA, C.; PAULAJÙNIOR, T. J.; BORÉM, A. (ed.). **Beans:** Aspectos Gerais e Cultura no Estado de Minas.

Viçosa: UFV, 1998. 596 p.

RAVEN, P.H., EVERT, R.F. & EICHORN, S.E. 2001. **Plant biology**. Guanabara-Koogan, Rio de Janeiro.

RODRIGUES, J. D.; FIOREZA, S. L.; Regulators are indispensable for many crops to achieve good levels. **Revista Visao Agricola**, Piracicaba, SP. N. 13, p. 35 - 39, 2015.

RUIZ, J. M. et al. Nitrogen metabolism in peper plants with different bioregulators.

Journal of Agricultural and Food Chemistry, Washington, v. 48, n. 7, p. 29252929, 2000.

SALVADOR, C.A. **Analysis of the** 2011/2012 **agricultural situation.** 2011

Available at: http://

www.agricultura.pr.gov.br/arquivos/File/deral/Feijao_2011_2012.pdf. Accessed at

April 17.

SANTOS, A.B.; FAGERIA, N.K.; SILVA, O.F. & MELO, M.L.B. Fava bean response to nitrogen management in tropical varzeas. **Pesq. Agropec. Bras.**, 38:12651271, 2003.

SANTOS, L. F. A. Behavior of common bean genotypes, of the carioca group, in the dry season, in Uberlândia-MG. 2010. 22 f. **Monografia** (Graduation in Agronomy), Universidade Federal de Uberlândia, Uberlândia. 2012.

SANTOS, M. A. N. Evaluation of common bean genotypes, of the carioca group, in the winter season, in Uberlândia-MG. 2012. 26 f. **Monograph** (Graduation in Agronomy), Universidade Federal de Uberlândia, Uberlândia. 2012.

SILVA, M. G.; ARF, O.; SA, M. E.; RODRIGUES, R. A. F.; BUZETTI, S. Soil management and nitrogen fertilization in winter bean crops. **Scientia Agricola**, Piracicaba, v. 61, n. 3, p. 307-312, 2004.

SILVA, T. R. B.; SORATTO, R. P.; CHIDI, S. N.; ARF, O.; SA, M. E.; BUZETTI, S. Doses and timing of nitrogen applications in winter bean cover crops. **Cultura Agronômica**, v. 9, n. 1, p. 1-17, 2000.

SILVEIRA, P. M.; DAMASCENO, M. A. Doses and installments of K and N in irrigated bean crops. **Pesquisa Agropecuària Brasileira**, v. 28, n. 11, p. 1269-1276, 1993.

SIQUEIRA, J.O.; MOREIRA, F.M.S.; GRISI, B.M.; HUNGRIA, M. & ARAUJO, R.S. **Microorganisms and soil biological processes**: An environmental perspective. Santo Antônio de Goiàs, Embrapa-CNPAF; Londrina, Embrapa-CNPSo; Brasilia, Embrapa- SPI, 1994. p.47-50. (Documents, 45)

SOUZA, S. R.; FERNANDES, M. S. Nitrogen. In: FERNANDES, M. S. **Nutriçao mineral de plantas**. Viçosa - MG: Brazilian Society of Soil Science, 2006. p. 216-252.

SRIVASTAVA, H. S. et al. Cytokinins affect the response of greening and green bean leaves to nitrogen dioxide and nutrients nitrate supply. **Journal of Plant Physiology**, Kusterdingen, v. 144, n. 2, p. 156-160, 1994.

STOLLER DO BRASIL. **Stimulate Mo in vegetables.** Cosmópolis: Stoller do Brasil. Divisâo Arbore, 1998. v. 1, 1 p. (Technical Information).

STRALIOTTO, R.; TEIXEIRA, M.G. & MERCANTE, F.M. Biological nitrogen fixation. In: AIDAR, H.; KLUTHCOUSKI, J. & STONE, L.F. **Produção de feijoeiro comum em vàrzeas tropicais.** Santo Antônio de Goiàs, Embrapa Arroz e Feijâo, 2002. p.122- 153

VELLINI, E. D.; ROSOLEM, C. A. **Agronomic efficiency of Stimulate.** Botucatu: UNESP, Dept. Agricultura e Melhoramento Vegetal, 1997. 8 p. (Technical Report).

VIEIRA, C. Mineral fertilization and liming. In: VIEIRA, C.; PAULA JUNIOR, T. J.; BORÉM, A. (Ed.). **Beans:** general aspects and cultivation in the state of Minas Gerais. Viçosa, MG: Universidade Federal de Viçosa, 1998. p. 123-152.

VIEIRA, C. **The common bean:** culture, diseases and improvement. Viçosa, MG: UFV, 220 p. 1967.

VIEIRA, E. L. **Action of biostimulants on seed germination, seedling vigor, root growth and productivity of soybean (*Glicine max*), bean (*Phaseolus vulgaris* L.) and rice (*Oryza sativa* L.).** 2001. 122 f. Thesis (Doctorate) - Luiz de Queiroz College of Agriculture, Federal University of São Paulo, Piracicaba, 2001.

VIEIRA, E. L.; CASTRO, P. R. C. Action of biostimulants on bean crops (*Phaseolus vulgaris* L.). In: FANCELLI, A. L.; DOURADO-NETO, D. (Ed.). **Irrigated beans:** technology and productivity. Piracicaba: ESALQ, 2003. p. 73-100.

VIEIRA, E. L.; CASTRO, P. R. C. Action of biostimulants on soybean (*Glycine max* (L.) Merrill). Cosmópolis: **Stoller do Brasil**, 2004.

VIEIRA, R. F.; TSAI, S. M.; TEIXEIRA, M. A. Nodulation and symbiotic nitrogen fixation in bean with native rhizobium strains, in soil treated with sewage sludge. **Pesquisa Agropecuària Brasileira**, Brasilia, v. 40, n. 10, p. 1047-1050, 2005.

VILHORDO, B.W. Morphology. In: ARAÙJO, R.S. (Coord). Common bean cultivation in Brazil. Piracicaba: **POTAFOS**, 1996. p.71-99.

VILLAS BÔAS, R. L, BULL, L. T., FERNANDEZ, D. Fertilizers in fertilization. In: FOLEGATTI, M. V. Fertilization: citrus, flowers, vegetables. Guaiba: **Agropecuària**, 1999. v. 1, p. 293-320.

WANDER, A. E. Bean production and consumption in Brazil, 1975 - 2005. **Economic information**. Sâo Paulo, v. 37, n° 2. 2007.

YOKOYAMA, L.P. Short-term aspects of bean production. In: AIDAR, H.;

KLUTHCOUSKI, J.; STONE, L.F. (Ed.). **Common bean production in tropical lowlands**. Santo Antônio de Goiàs: Embrapa Arroz e Feijâo, 2002. p.249-292.

9. ANNEXES

FIGURE 3: Collection of soil samples (0-20 cm).

FIGURE 4 Final composite soil sample.

	pH Água	P-rem	Mat Org	P	K	Ca	Mg	Al	H + Al	SB	CTC(t)	CTC(T)	m	V	B	Cu	Fe	Mn	Zn	S	Argila	Silte	Areia	COT
AMOSTRA 61	6,26	4,00		21,97	209,00	1,92	0,65	0,02	2,15	3,11	3,13	5,26	0,54	59,21										

FIGURE 5: Chemical analysis of the soil.

FIGURE 6. General view of the experimental area after sowing.

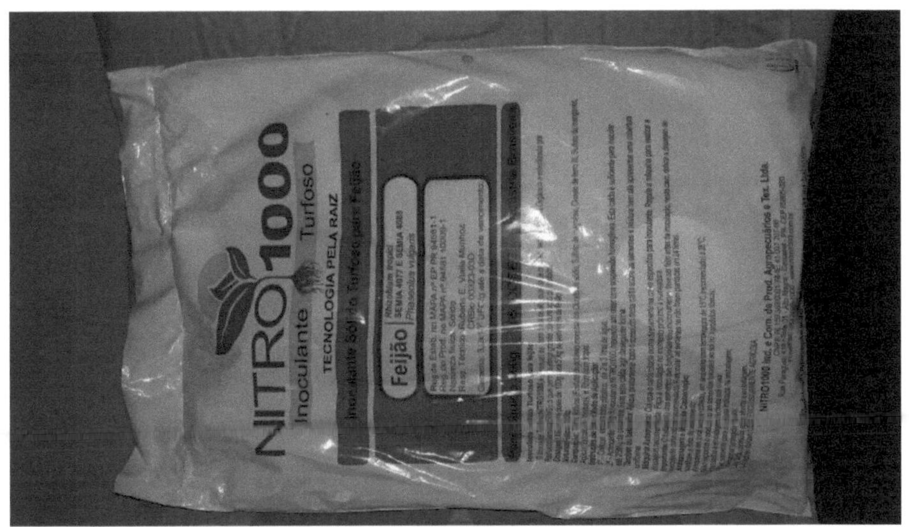

FIGURE 7 Peat inoculant used.

FIGURE 8 - First nitrogen top dressing (20 DAE) in the treatments (T1 and T2) that received mineral nitrogen.

FIGURE 9 - Second nitrogen top dressing (30 DAE) in treatments (T1 and T2) that received mineral nitrogen.

FIGURE 10. General view of the area (28 DAE).

FIGURE 11. General view of the area at reproductive stage r6 - flowering (first flower open).

FIGURE 12. Collection of plants with pá for evaluation.

FIGURE 13. Bean root system at the time of collection.

FIGURE 14. Weighing the dry mass of the aerial part of the treatments.

FIGURE 15. Root dry mass weighing.

FIGURE 16. Counting the number of pods per plant.

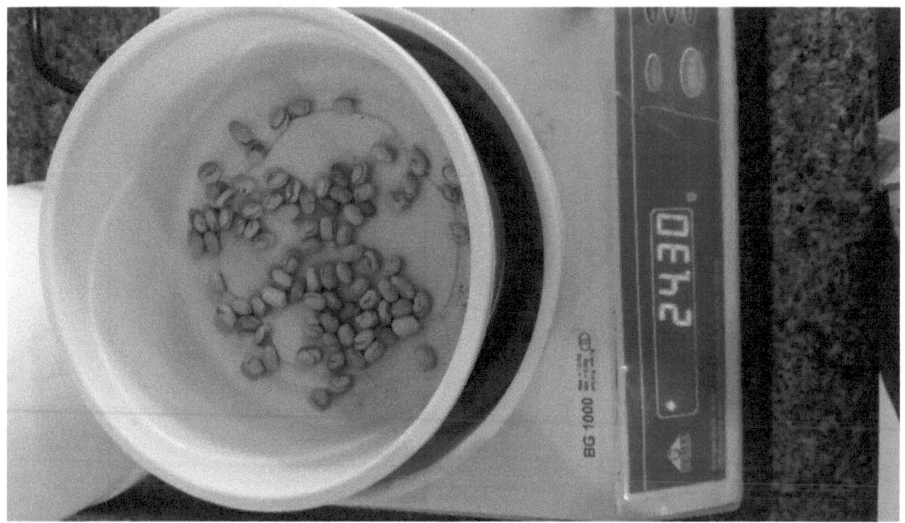

FIGURE 17. Weighing 100 grains.

I want morebooks!

Buy your books fast and straightforward online - at one of world's fastest growing online book stores! Environmentally sound due to Print-on-Demand technologies.

Buy your books online at
www.morebooks.shop

Kaufen Sie Ihre Bücher schnell und unkompliziert online – auf einer der am schnellsten wachsenden Buchhandelsplattformen weltweit! Dank Print-On-Demand umwelt- und ressourcenschonend produziert.

Bücher schneller online kaufen
www.morebooks.shop

Printed by Books on Demand GmbH, Norderstedt / Germany